I0483246

MRS. BUMBLEBEE

AND HER

COUNTING

WINGS

WRITTEN AND ILLUSTRATED
BY
GWENDOLYN BLACKSHEAR

Copyright © 2016 Gwendolyn Blackshear

All rights reserved.

No part of this book may be reproduced, stored in a retrieval system, or transmitted in any form or by any means, electronic, mechanical, photocopying, recording, or otherwise, without the prior written permission of the publisher, author and illustrator, except as provided by U.S.A. copyright law.

Published by Gwendolyn Blackshear, October 2016

Printed in the United States of America

Library of Congress Cataloging in Publication Data

ISBN -13: 978-1535558464

ISBN-10: 1535558466

Blackshear, Gwendolyn

Mrs. Bumblebee and her counting wings.

To:

all the awesome children I have met
who were always *ready to learn*!
Today, that child is *you*!

To:

London Coleman – my youngest consultant.

Mrs. Bumblebee loves to count.

She just landed in your town with her

counting wings! Let's count! Here

we go and here they come!

One music note

1
one

1
one

2

Mrs. Bumblebee

and her counting wings.

2

two

Two tea cups

1   2

one   two

Mrs. Bumblebee and her

counting, counting wings.

Three summer hats

three

1   2   3

one   two   three

6

Mrs. Bumblebee

and her counting, counting, counting wings.

4

four

Four flowers

1 **2** **3** **4**

one two three four

8

Mrs. Bumblebee and her counting, counting, counting, counting wings.

Five hearts

5

five

Mrs. Bumblebee and her
counting, counting, counting,
counting, counting wings.

6

six

Six apples

1 2 3 4 5 6

one two three four five six

12

Mrs. Bumblebee

and her counting, counting, counting, counting, counting, counting wings.

Seven honey combs

7

seven

1   2   3   4   5   6   7

one   two   three   four   five   six   seven

14

Mrs. Bumblebee

and her counting, counting, counting,

counting, counting, counting, counting

wings.

Eight cupcakes

8

eight

1 2 3 4 5 6 7 8

one two three four five six seven eight

Mrs. Bumblebee

and her

**counting, counting, counting,
counting, counting, counting,
counting, counting wings.**

Nine ice cream cones

9

nine

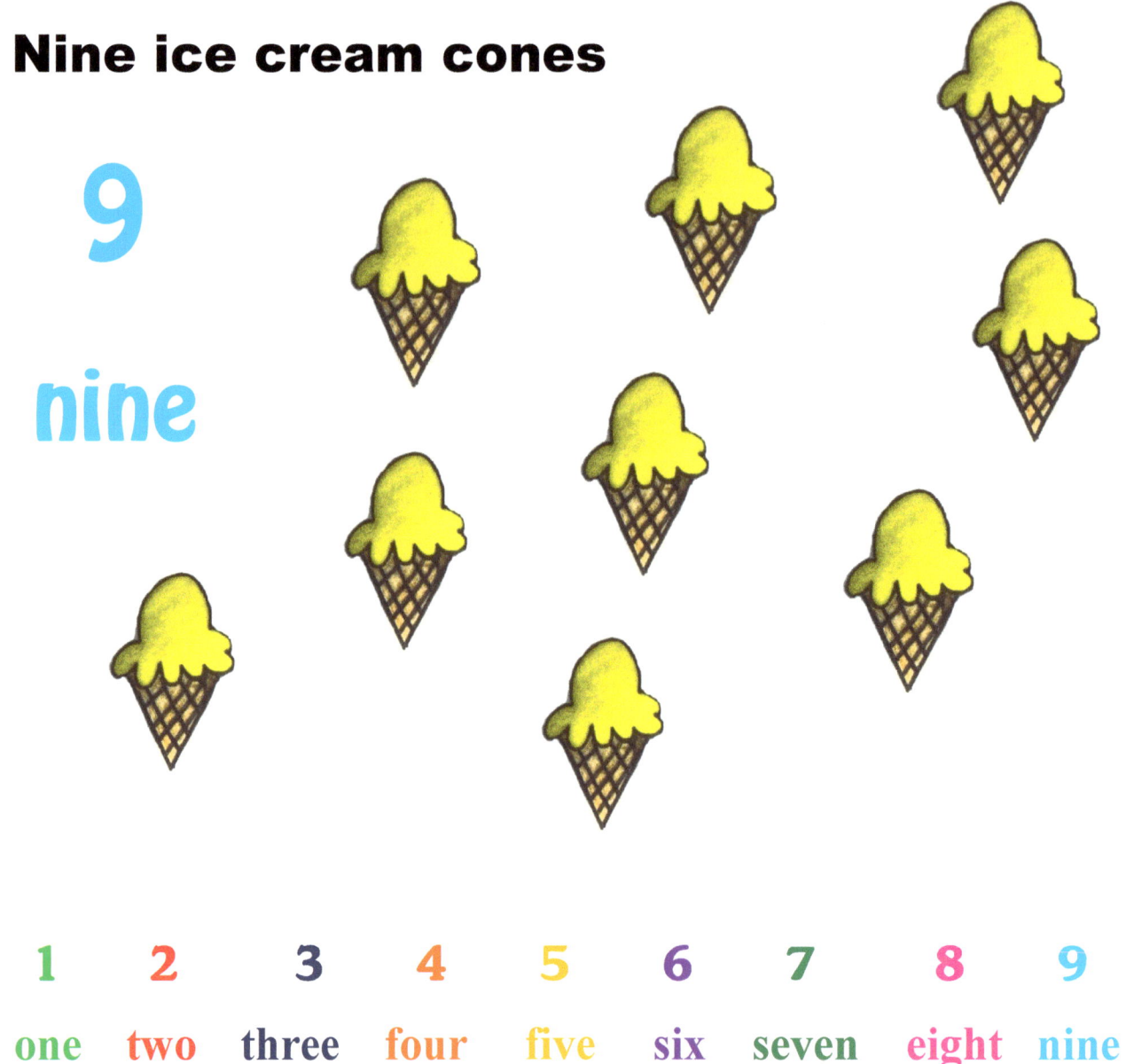

1 2 3 4 5 6 7 8 9

one two three four five six seven eight nine

Mrs. Bumblebee

and her counting, counting, counting, counting, counting, counting, counting, counting, counting

wings.

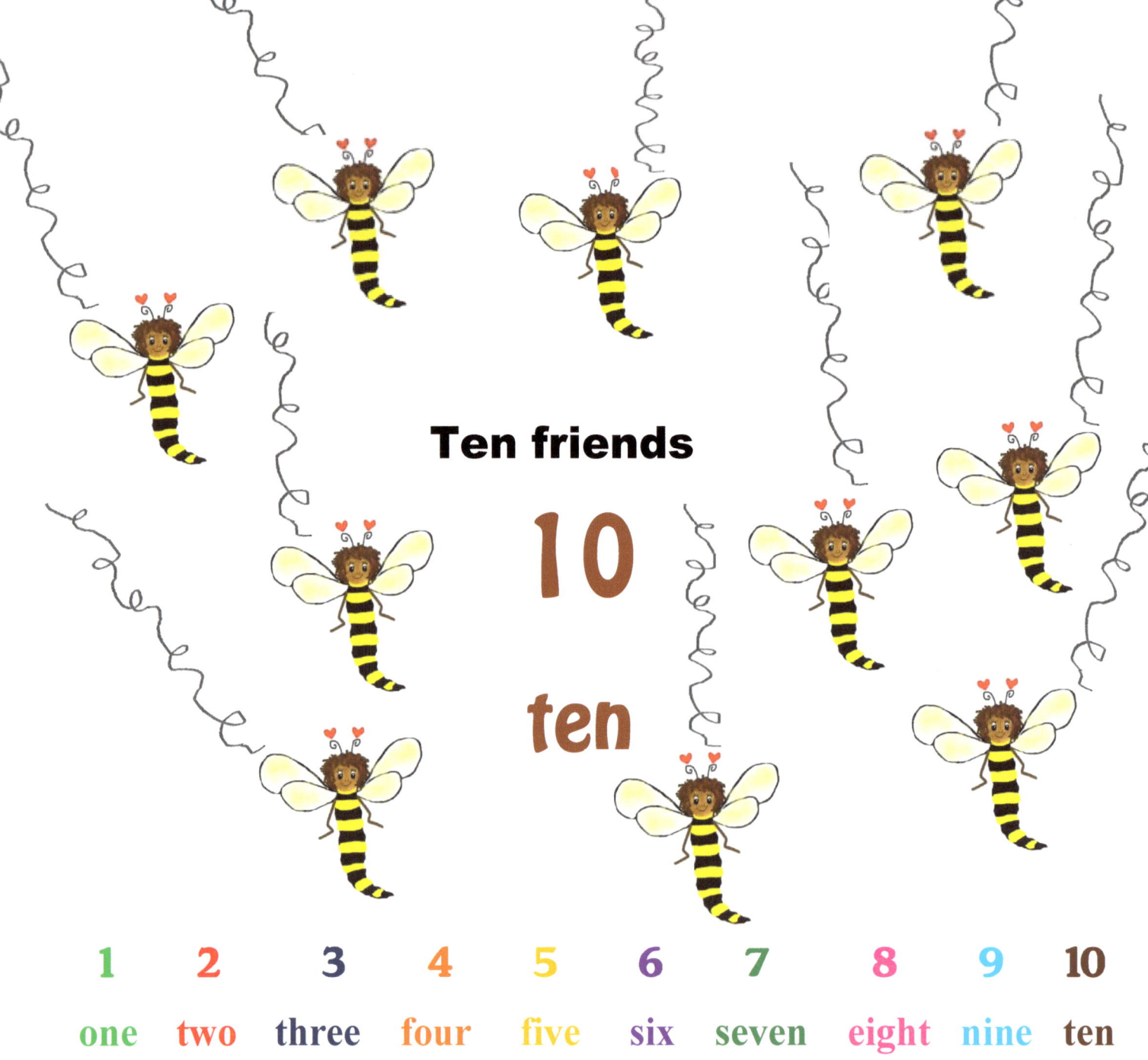

Ten friends

10

ten

1   2   3   4   5   6   7   8   9   10

one   two   three   four   five   six   seven   eight   nine   ten

Mrs. Bumblebee and her

counting, counting, counting, counting,

counting, counting, counting, counting, counting, counting

wings.

Here we go, and here they come!

Mrs. Bumblebee

and her counting wings, and her counting friends.

Today, that friend is *you!*

The story behind Mrs. Bumblebee…

As the bell rang on Mrs. Blackshear's first day welcoming in a new class of students, she said, "Good morning boys and girls, my name is Mrs. B! Welcome to class! Are you ready to learn with me today?" A kindergartener with the brightest eyes, and one silent tear, came up to Mrs. Blackshear and said, "Good morning Mrs. Bumblebee! Is it almost time for my Mommy to pick me up"? Mrs. Bumblebee replied with a smile, "It won't be very long dear. We will learn a few fun things today and she will be on her way!" The student jumped up and down, offered a big smile and said, "Okay, I'm ready learn!", as she took her place in her assigned seat.

That day, and at every setting that followed,

this teacher was known by all

as

Mrs. Bumblebee

The Author:

Gwendolyn Blackshear is excited to be debuting as a writer and illustrator with this book. She was so inspired by so many young learners she met and taught, she decided to share in the teaching process, even more. She also sings, blends her own tea, hosts tea parties, clowns around town and has a crush on the Easter Bunny. She is the mother of two very kind adults and five super awesome grandchildren. She is an education and nonprofit consultant and speaker for schools, church and community organizations. Gwendolyn is currently pursuing a doctorate in education. She lives, works, worships and plays in northeastern Ohio with her famous puppy Morocco.

Teachers and Parents:

First, thank you for all you do! I would also like to thank you for giving me this opportunity to inspire you little stars. Please feel free to check out my website for the *Mrs. Bumblebee and her Counting Wings* **color and practice sheets, and more!**

Stars come from stars,

Gwendolyn

For visits from Gwendolyn and/or book orders, contact her at:

EMAIL: starscomefromstarslearningrock@gmail.com

WEBSITE: starscomefromstarslearningrock.com

PHONE: (330) 794-3163

www.ingramcontent.com/pod-product-compliance
Lightning Source LLC
Chambersburg PA
CBHW050424180526
45159CB00005B/2396